Bibliographic information published by the German National Library:

The German National Library lists this publication in the National Bibliography; detailed bibliographic data are available on the Internet at http://dnb.dnb.de .

Imprint:

Copyright © 2016 GRIN Verlag, Open Publishing GmbH
Print and binding: Books on Demand GmbH, Norderstedt Germany
ISBN: 9783668302235

This book at GRIN:

http://www.grin.com/en/e-book/340402/3d-printing-with-biopolymers-on-textile-knitted-structures

Aliea Mohammad

3D Printing with Biopolymers on Textile Knitted Structures

GRIN Publishing

3D Printing with biopolymers on knitted structures

By Aliea Mohammad

Table of contents

Table of figures

1. The principle of 3D printing technology

Additive manufacturing or 3D printing is the process of turning digital files into physical, three dimensional objects. This is realized using additive processes, which imply successfully laying down very thin layers of material until the object is finalized.

Since invented, it has been used for the purpose of rapid prototyping, and has evolved into a next generation manufacturing technology with the potential of allowing rapid, on site and on demand production of parts and end-products, signaling the beginning of a third industrial revolution. [1]

The aim of this research is to successfully print different three dimensional structures on a variety of knitted fabrics, in order to observe the properties and the behavior of the biopolymers in these circumstances. To do so, 3 separate objects in form of thin rectangles were 3D printed on 7 different surfaces, which were later subject of a peel test that measured the adherence of the polymer to the knitted structure.

Even though 3D printing can only occur by means of additive processes, what may differ is the way layers are built to create the final object, resulting in several printing technologies. These have been classified in 2010 by the American Society for Testing and Materials (ASTM) group F42-Additive Manufacturing into 7 categories according to the Standard Terminology for Additive Manufacturing Technologies, as follows:

1. Material Extrusion
2. Material Jetting
3. Binder Jetting
4. Vat Photopolymerisation
5. Powder Bed Fusion
6. Sheet Lamination
7. Directed Energy Deposition

During the following experiments the material extrusion method was used, otherwise known as Fuse Deposition Modeling (FDM), which works with plastic filament or metal wire unwound from a coil that supplies material to an extrusion nozzle. The nozzle can regulate the flow and is able to move both vertically and horizontally, directly controlled by a computer-aided manufacturing (CAM) software package. It is also heated up to very high temperatures, so that the material can be melted and extruded in liquid form, and then harden back immediately after the extrusion. Nevertheless, for more intricate designs, an additional support is required for maintaining the shape and steadiness of the object in formation.

---

[1] Cf. http://3dprinting.com/what-is-3d-printing/

FDM is a commonly used technique which differentiates itself from the others by the fact that the material is added under constant pressure and continuous stream that must be kept steady and maintain its speed to enable accurate results. Also, the quality of the end product is reduced because of the nozzle radius, and accuracy and speed are low when compared to other processes. [2]

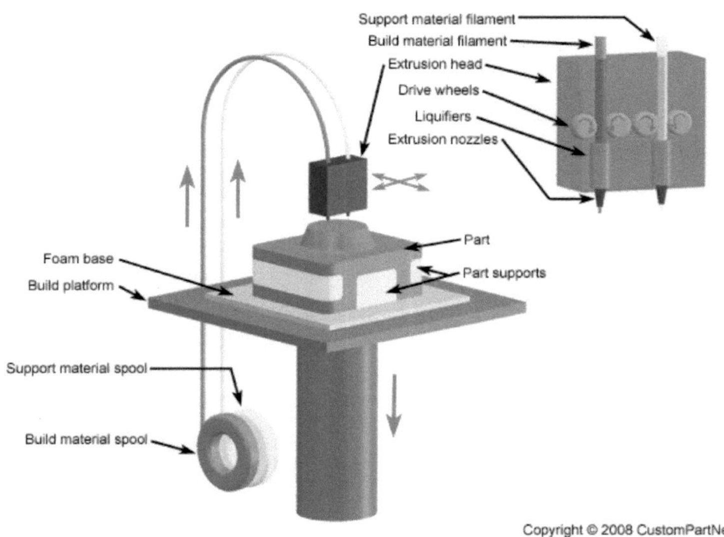

Figure 1 Fused deposition modeling sketch by CustomPartNet

---

[2] Cf. Christopher Barnatt at http://explainingthefuture.com/3dprinting.html

2. Fabrics

The first step of this research was to procure the necessary materials on which to carry the experiments. The requirements for the fabrics were for them to be knitted structures but differ in composition or type of inter-looping, and to have different properties such as thickness, fineness or coarseness, so that the results would be comparable. After a stage of filtering, 6 fabrics were chosen, half of which made out of natural fibers, two synthetic, and one double faced.

The adherence between a textile surface and a polymer is strongly related to the structure and physical properties of the material in use, therefore the printer needs to be set accordingly, in order to enable the best results. That being said, the samples were closely analyzed to obtain information about the thickness, fineness, and microscopic appearance of the fabric, as well as fiber thickness and loop size.

The fabric thickness tester was used for measuring the thickness, whereas the fiber thickness was determined via microscope, using the option "measure distance between two points". In the following table, a clear overlook of the composition, inter looping method and characteristics of the fabrics is given.

Table 1 Physical properties of fabrics

Yarn	Type	Fabric thickness	Fiber thickness	Loop length	Loop width	Loop diameter	Wale density	Course density
Cotton	Fleece	30 mm	17.8 um	181 um	136 um	227 um	22 w/cm	26 c/cm
Cotton	Pique	13 mm	17.3 um	260 um	195 um	325 um	24 w/cm	20 c/cm
Cotton	Single Jersey	22 mm	12.6 um	216 um	157 um	267 um	20 w/cm	36 c/cm
Cotton	Double knit	25 mm	18.3 um	193 um	125 um	230 um	36 w/cm	42 c/cm
Polyamide	Double knit	25 mm	18.6 um	193 um	125 um	230 um	43 w/cm	56 c/cm
Polyester	Jacquard	23 mm	30.4 um	251 um	189 um	314 um	18 w/cm	22 c/cm
Polyester	Warp	19 mm	18.4 um	237 um	185 um	301 um	22 w/cm	30 c/cm

3. Microscopic view

3.1 Cotton Single Jersey

Magnification	Microscopic view	Physical properties
30x	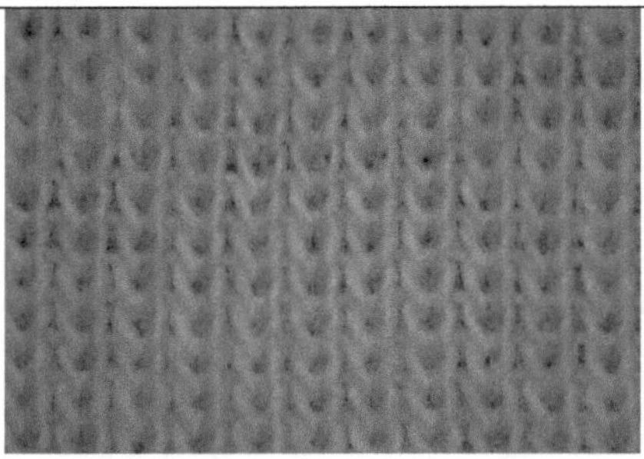	<ul><li>Face side and back side of fabric are different</li><li>Roll of fabric occurs at the edges</li><li>Wales are clearly visible on the face side of the fabric</li><li>Extensibility in widthwise is approximately twice than length.</li><li>Thickness of fabric is approximately twice the diameter of yarn used.</li></ul>
200x 3D	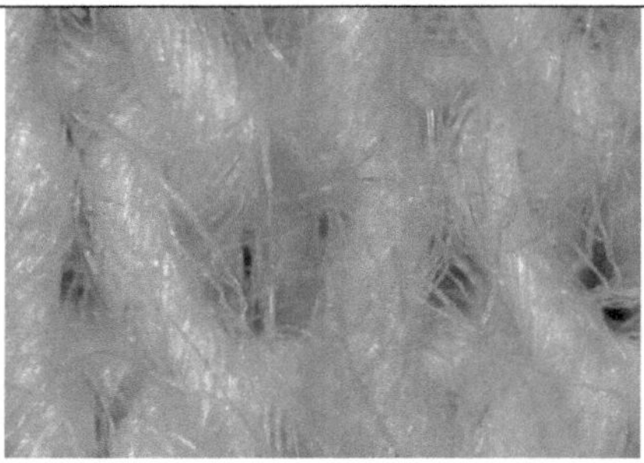	
50x	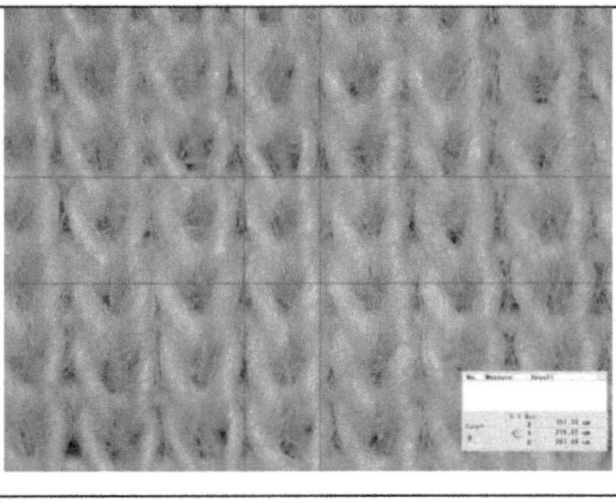	

Table 2 Cotton single jersey

3.2 Cotton Fleece

Magnification	Microscopic view	Physical properties
30x	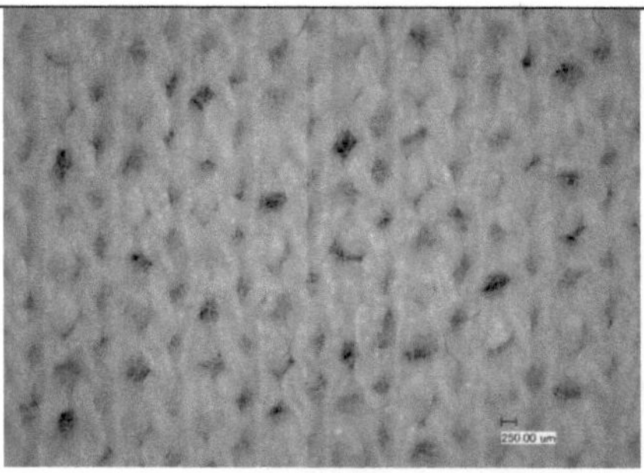	• Features soft piles of yarn on the inside • Very good moisture absorption • Good stretch • Edges roll only slightly • Insulates well without being too heavy • Face side and back side of fabric are different • Loops are clearly visible on backside
200x/ 3D	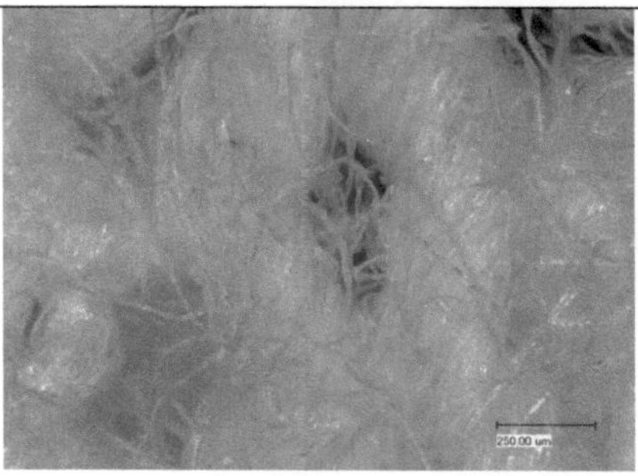	
30x	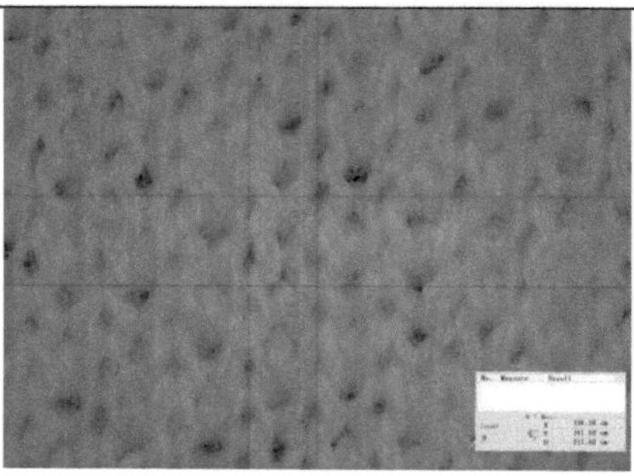	

Table 3 Cotton fleece

3.3 Cotton Pique

Magnification	Microscopic view	Physical properties
30x		• Stiff, durable rib fabric • Embossed pattern produced by a double warp thread • Thick and medium weight • Good breathability • Good abrasion resistance • Face side and back side of fabric are different • S twist of the yarn is noticeable in the 200x magnification
200x/ 3D		
50x		

Table 4 Cotton Pique

3.4   Double knit cotton side

Magnification	Microscopic view	Physical properties
30x		• Two knitted layers connected by a single thread • Double thickness • The two fabrics have very contrasting properties • Right side on both faces
200x/ 3D		
50x		

Table 5 Double knit cotton side

3.5 Double knit polyamide side

Magnification	Microscopic view	Physical properties
30x		• Two knitted layers connected by a single thread • Double thickness • The two fabrics have very contrasting properties • Right side on both faces • Melting point of Polyamide is 220 degrees Celsius
200x/ 3D	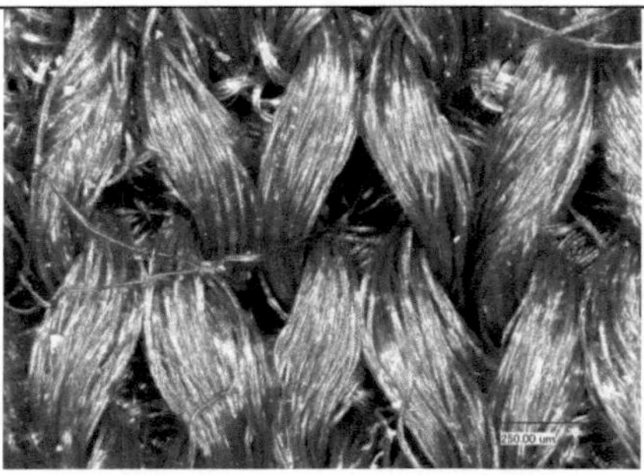	
50x	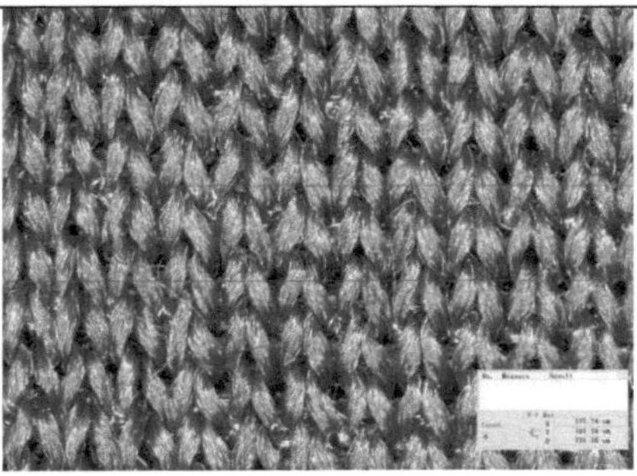	

Table 6 Double knit polyamide side

3.6 Polyester jacquard

Magnification	Microscopic view	Physical properties
30x	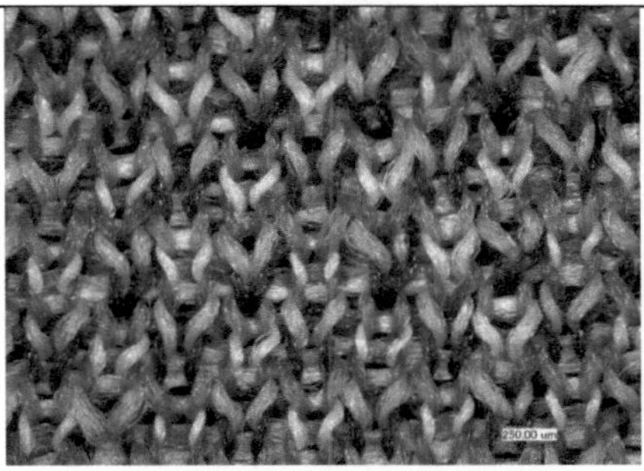	• Five color Jacquard • Face side and back side of fabric are different • Bird's eye backing • Edges roll strongly • Rib based, double jersey weft knit • Unlimited pattern possibilities • Melting point of Polyester is 250 degrees Celsius
200x/ 3D	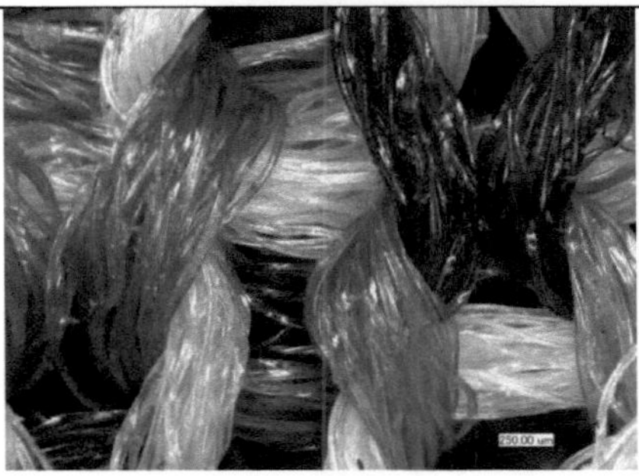	
50x		

Table 7 Jacquard polyester

3.7 Polyester warp

Magnification	Microscopic view	Physical properties
30x	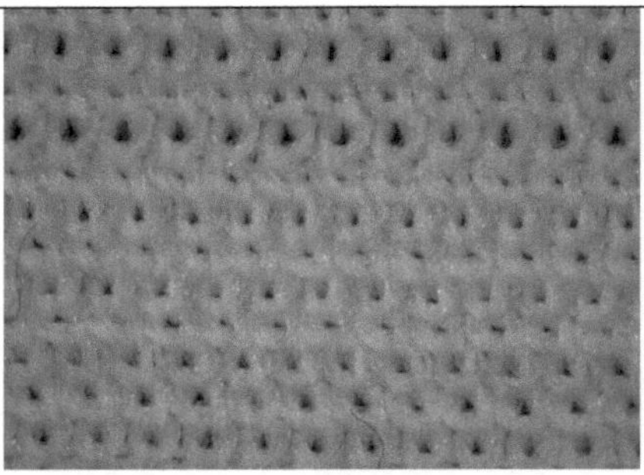	• Large working widths possible • Low stress rate on the yarn • Thin but very strong • Face side and back side of fabric are different • Fine lengthwise ribs are visible on the face side • Melting point of Polyester is 250 degrees Celsius
200x/ 3D	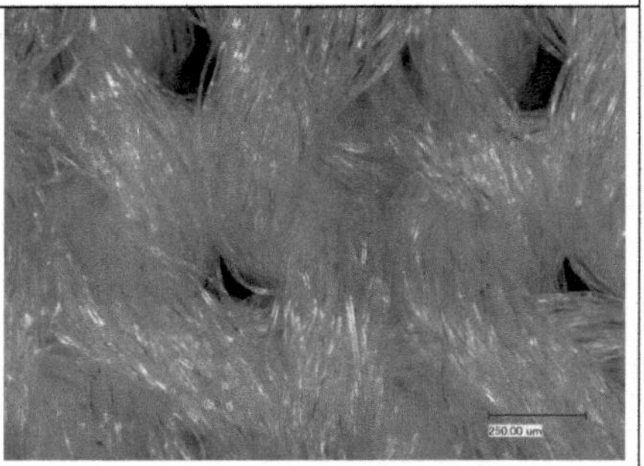	
50x	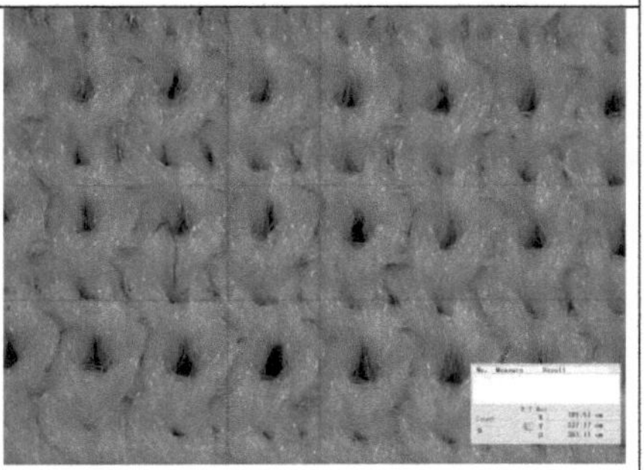	

Table 8 Warp polyester

As a result of the analysis, it has been observed that the physical structure of the chosen fabrics is very different from one another, thus presenting a big advantage in contrasting and comparing the results of the post-printing tests.

In the 30x magnification view, the type of inter-looping of the knit was observed, whereas in the 200x view, the filament or fibers can be seen in a very clear way, making it easy to measure their width. A 3 dimensional view was necessary for the 200x magnification, so that the fibers throughout the entire depth of the materials could be visible, and give a better overview of how the knitted structure is layered. The perfect magnification for measuring the loop width and length, course density and wale density was the 50x one. Here the loops, courses and wales can easily be counted, and the hairiness of the fabric can also be observed.

4. CAD pattern development

The first step into creating a physical object by 3D printing is to make a virtual design of it. This is done in a Computer Aided Design (CAD) file using a 3D modeling software, which slices the final model into hundreds or thousands of horizontal layers in order to prepare the digital file for printing. Once the sliced file is uploaded in the 3D printer, the object can be created as the printer reads every 2D image and develops the object, blending each layer so that they are barely recognizable in the final result.

As mentioned in the beginning, the adherence of the biopolymer to the fabric needs to be measured by means of a peel test, which requires a certain size and form of the objects to be tested; more specifically, the 3D printed object must have a similar appearance to that of a textile fabric. To achieve this, a thin, rectangular digital model was created with the help of Inventor®, a software for mechanical construction and 3D CAD development. The virtual design is 150 mm long, 30 mm wide and is composed of 2 layers of biopolymer, each having a thickness of 0.2mm.

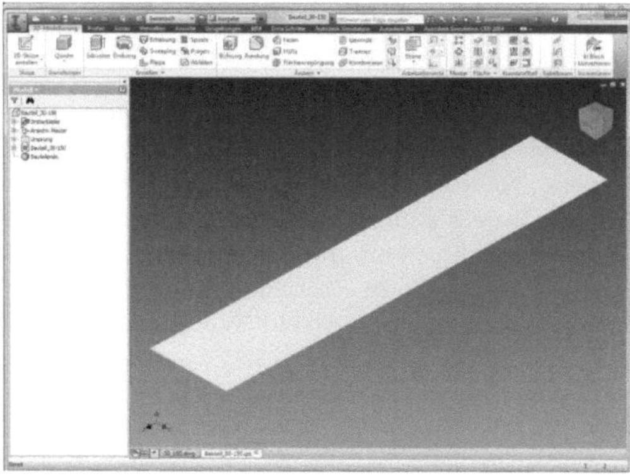

Figure 2 CAD Model

5. Polylactic acid

PLA is a biodegradable thermoplastic aliphatic polyester, that differs from others in that it is derived from renewable sources such as corn starch, sugar cane, tapioca roots or chips. Although the name is potentially ambiguous, PLA is not a polyacid, but rather a polyester. It is relatively cost efficient to produce and has, accordingly, the second largest production of any bioplastic, with a wide variety of applications including fiber production, decomposable packaging material, biodegradable medical implants, injection molding, as well as a feedstock material in desktop fused filament fabrication-based 3D printers. It is one of the two most commonly used desktop 3D printing materials next to Acrylonitrile Butadiene Styrene (ABS). [3]

Its popularity is also due to the fact that PLA represents a long awaited innovation of cost efficient, non petroleum plastic production, with the major advantage of being extremely versatile and naturally degradable when exposed to the environment. As a relevant comparison, it would typically take a PLA bottle left in the ocean approximately 6 to 24 months to completely degrade. On the other hand, placed in the same environment, it would take the same bottle made out of conventional plastics several hundred to a thousand years to degrade, which leads to a great leap forward in the manufacturing world. For this reason, PLA has a high potential in the manufacturing industry of short lifespan applications such as packaging, disposables, and medical uses. However, despite its highly appraised ability of degrading when exposed to certain elements over a long time, PLA is extremely robust and withholds its structure in any normal application such as plastic electronics parts. [4]

However, within this research the printing is done on knitted structures, which are extremely flexible and would most probably end up in contact with the human body where a lot of freedom of movement is needed. For this reason, the filament that is being used must match the properties of the fabrics, otherwise there is a strong chance that the printed model will be damaged because of its inability to mold to fit its environment. Therefore a more flexible material called soft PLA was used, which can bend or flex and feels much like rubber. Soft PLA must be printed at a lower speed and higher temperature compared to regular PLA (220°-240° as opposed to 180°-220°), which makes it more difficult to work with. Nevertheless, Soft PLA has a 50% lower carbon footprint than traditional 3D printing materials, which makes it perfect for mass production as well as prototyping.

---

[3] https://www.matterhackers.com/3d-printer-filament-compare
[4] Cf. Tony Rogers https://www.creativemechanisms.com/blog/learn-about-polylactic-acid-pla-prototypes

6. Printing process

Next, the digital file is inserted into the 3D printing software Orcabot Repetier-Host, where the number of layers can be adjusted and the object can be multiplied and set at the desired angle onto the printing bed. The previously mentioned rectangle was replicated two times in the exact same dimensions, so that the printing accuracy in different points of the printing bed could be observed.

To enable constant printing conditions, a configurated profile has been created, where the printer, filament and printing settings can be adjusted and saved for further use. In the following attempts, the standard preconditions for the objects were to be printed in 2 layers each, with a distance from the nozzle to the printing bed depending on the fabric thickness, mostly of 0.0 cm. Temperature also represents a crucial factor for achieving optimal results, especially as each polymer has a different melting point and distinct designs require slightly different resin flow rates in order to achieve the desired appearance. Furthermore, when printing directly on textile surfaces, it has to be kept in mind that synthetic materials have different melting points, which if reached by the nozzle temperature, might cause the fabric to melt or change its appearance, since usually the extruder gets to be in direct contact with the textile surface. In this case, the optimal temperature for PLA soft is 240°,the printing bed was set at 60°, and the feed rate and flow rate at 100%.

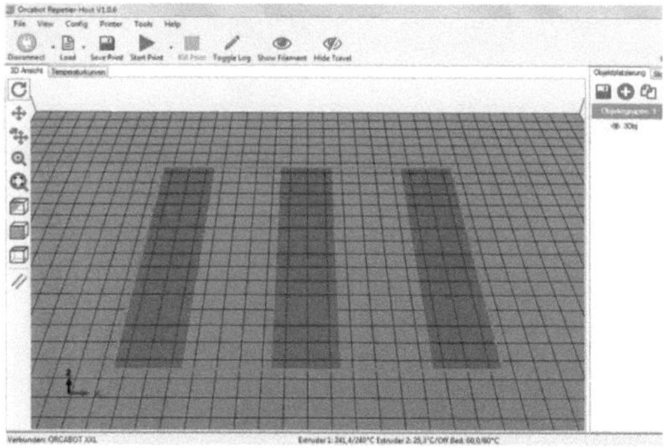

Figure 3 CAM printing model

As preparation for the printing, each fabric sample was cut out in even parts smaller than the printing bed, and fixed on every side by means of tape. The fabric was slightly but evenly stretched in all directions so that the printing surface could be even and the nozzle wouldn't be able to move the fabric when passing over it. Additionally, a thin piece of tape was applied at the bottom of the area to be printed, so that a small part at the beginning of the rectangles can be left unattached to the fabric. The printed part on this piece of tape can easily be detached after the printing process, and can later on be fixed into the peel test machine, for testing the adherence.

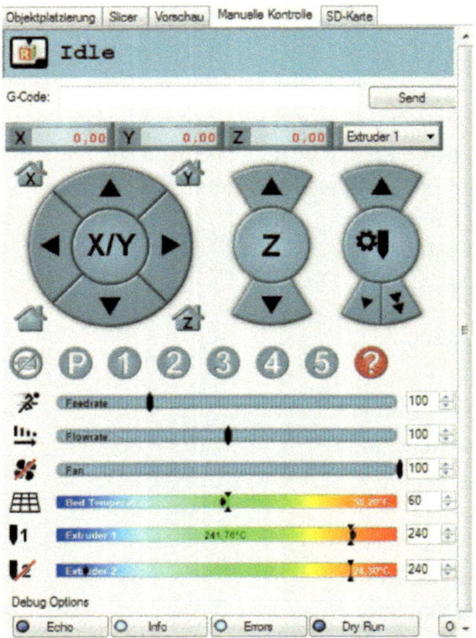

Figure 4 Printing parameters

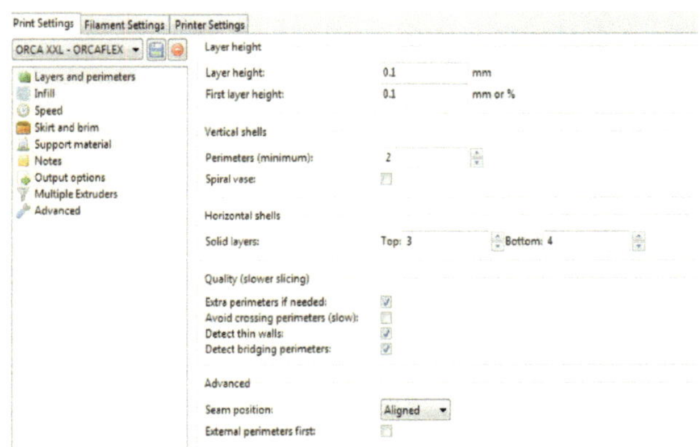

Figure 5 Print settings

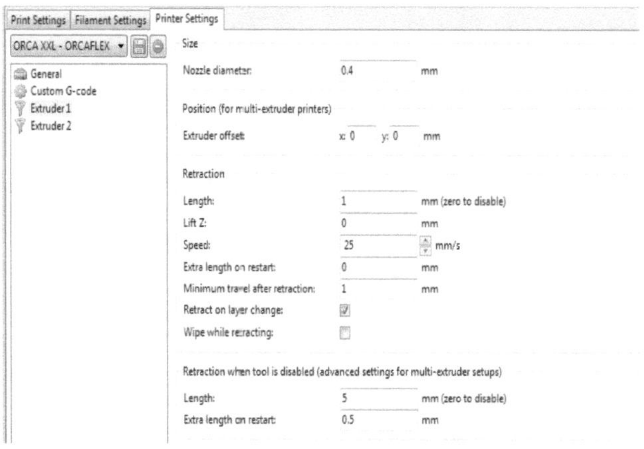

Figure 6 Printer settings

6.1 Pique

The first series of trials was carried out on the 100% cotton Pique knit, for which two failed attempts were needed in order to find out the right printing settings to match the surface and thickness of the fabric. At first, the resin wasn't properly extruded from the nozzle and not enough PLA reached the fabric. However, after these issues were fixed, the printing was very successful on all the three specimens. For this first series, the fabric was cut out in small pieces and only one rectangle at a time was printed in the same spot in the center of the printing bed. Additionally, one of the three samples was placed in a different direction that the other two, so that two samples were printed lengthwise and one was printed widthwise. Although this is an interesting matter to be observed when testing the adherence, a different printing method was needed for enabling constant results, and lowering the time needed for changing and taping each sample on the machine. Accordingly, for the next 7 series the fabrics were cut out almost the same size as the printing bed and all three rectangles were printed in one go with a distance of approximately 2 cm between each other. This way, not only could all three models be printed in the same circumstances, but also a comparison between different points on the printing bed could be made later on.

6.2   Single jersey

Next, the 100% cotton single jersey knit was printed on, but the first attempt failed because of the nozzle temperature, which was set as 220 degrees Celsius. Even though in the instructions for the soft PLA filament it is stated that the optimal temperature lays between 220 and 240 degrees, it seems that 220 degrees is extremely low, since the results were very bad. For this reason 240 degrees was used again from here on out, since this seemed to be the perfect temperature for this type of filament, and the results on the single jersey sample turned out to be very good at this temperature.

6.3 Fleece

The most challenging material to work with turned out to be the 100% cotton fleece. In the first try it was noticed that because of the high thickness of the material, a z=0 mm distance from the nozzle was too low, and the resin wasn't able to properly extrude from the nozzle because of the lack of space. The result was a very blurry and uneven surface on the textile, whereas on the green tape that was used to separate the ends of the rectangle, there was barely any extrusion taking place since that spot was even higher because of the addition of the tape. The thin layer of PLA at the ends made it impossible to remove the green tape without ripping the rectangle up. The next attempt was to take a z=0.2 mm distance from the nozzle, to allow proper flow of the material and to use a different, much thinner tape at the ends of the model. Unfortunately, two attempts of this sort failed: the first, because of the poor flow of material and the lack thereof on the first layer of the strip. In the second trial of this kind, even though the appearance of the PLA on the textile itself was good, the strip was too weak at the ends to bear the separation from the tape and therefore broke. The solution was to increase the z distance to 0.3 mm and to print 3 layers on each rectangle for a stronger grip when it comes to taking the tape off at the ends. The final try using these parameters was successful.

6.4   Double face cotton

As the fourth fabric was a double faced knitted structure, which is formed by one layer of polyamide and one layer of cotton knit adjoined in the middle by a thread, the best way to go about it was to observe how both sides reacted to the printing process. First, the cotton side was very easy to work with, as the structure of the knit was very fine and the fabric wasn't very thick. The results were satisfactory from the very first try, printing with z=0 distance from the nozzle and in two layers thickness.

6.5 Double faced polyamide

However, the side composed out of 100% polyamide did not behave that well under these printing circumstances. Seeing as the actual melting point of polyamide is usually 220 degrees Celsius, it is understandable that the surface of the fabric was severely altered upon contact with the 240 degrees heated nozzle. A layer of black melt was indeed noticeable even with the naked eye, but regardless of that, the printing process took place in a normal manner, as it did on the cotton side.

6.6   Weft polyester

Moving on with the synthetic fibers, the next trial was made on a 100% polyester weft knitted fabric. Since the melting temperature of PES is over 250 degrees Celsius, no problems occurred such as in the case of the polyamide, and the nozzle temperature did not damage the fibers at all. The printing went on very smoothly at the first strip, but some extrusion problems occurred during the printing of the second layer at the other two strips. More specifically, the nozzle was repeatedly clogged with material and the flow was damaged, causing unevenness and lack of material in some points.

6.7   Jacquard polyester

The first trial (nozzle distance of 0.0 mm) failed because the nozzle did not have enough space to extrude the material especially onto the tape at the end of the strips, since that is the highest point on the textile. The Jacquard knit with bird's eye backing is a thick and bulky material like the cotton fleece, and must therefore also be printed on with a z nozzle distance of 0.2 mm. However, during the printing process it was observed that the fabric still contained certain detergents and oils on it, so the fabric was washed and printed on again. This time, the printing process succeeded even with a z distance of 0.0 mm, showing that the fabric surface became much more printing friendly after washing. Both of the specimens are later tested and the results pre and post washing are compared.

Figure 7 Printing center strip on double knit CO side

7. Separation force test

After successfully printing the desired models on every sample, the next relevant step was to test the adherence of the PLA to the fabrics in order to enable an exact comparison between them. This was done in the quality control laboratory using a Zwick 1455 tensile strength testing machine. The samples were cut to the exact length and width as the PLA strip and placed on the machine in the following way: using only the side of the specimens where the two layers are separate, the textile part was clamped to the lower part of the stretching device and the PLA strip was clamped to the upper part, leaving the adhered part free. The peel test was executed according to the DIN 53357 standard.

The three specimens of each fabric were tested and the results measured in cN are exemplified in the following graphs. Since all three models were printed in one go, the specimen noted with the number 1 represents the strip that was printed on the right side of the printing bed, the specimen 2 represents the strip on the middle and the specimen number 3 represents the strip on the left. Furthermore, since the strips were partially clamped during the procedure, a buffer was needed regarding the length of the tested area. Therefore, only 110 mm out of the 150 mm length of each strip were taken into consideration: 20 mm were disregarded at each end of the strips.

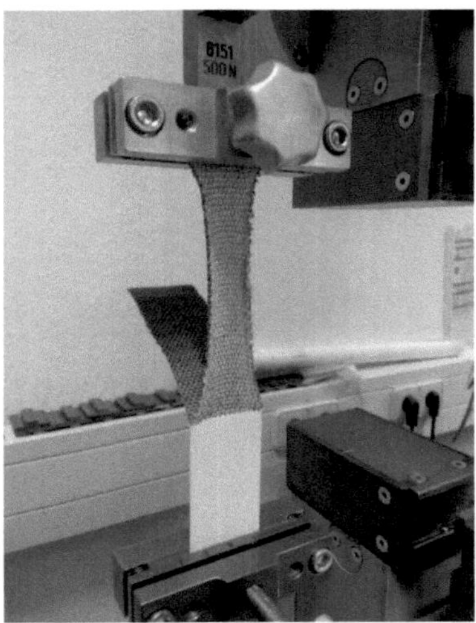

Figure 8 Jacquard PES peel test

7.1 Pique

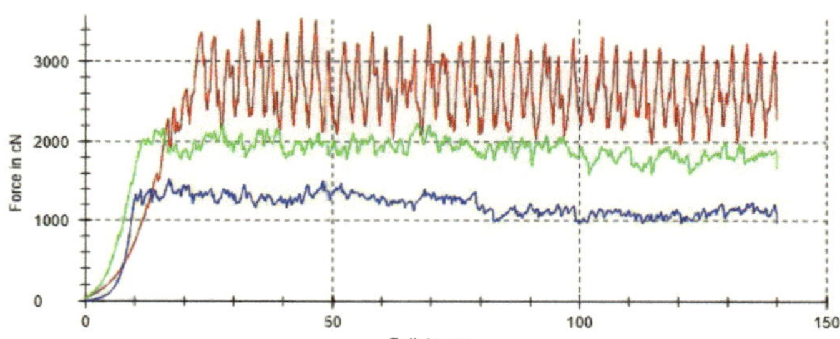

Figure 9 Pique graph

Legend	Nr	Index	Adhesion test Single value cN	Adhesion test cN
	1	1	2564	2481
		2	2587	
		3	2631	
		4	2305	
		5	2358	
		6	2439	
	2	1	1875	1885
		2	1925	
		3	1961	
		4	1847	
		5	1847	
		6	1857	
	3	1	1423	1193
		2	1216	
		3	1299	
		4	1156	
		5	973,3	
		6	1088	

Table 9 Pique values

As mentioned before, the first specimen cf this series was printed widthwise as opposed to the other two that were printed lengthwise. This explains the big difference between the results and also gives an indication on how the direction of the knitted material on the printing bed plays a huge role in the adhesion results. Also, it is to be noticed that the strip that was printed in the center of the printing bed (strip number 2), which is also the strip that is printed first before the other two, has a higher adhesion value that the strip on the left..

7.2 Single Jersey

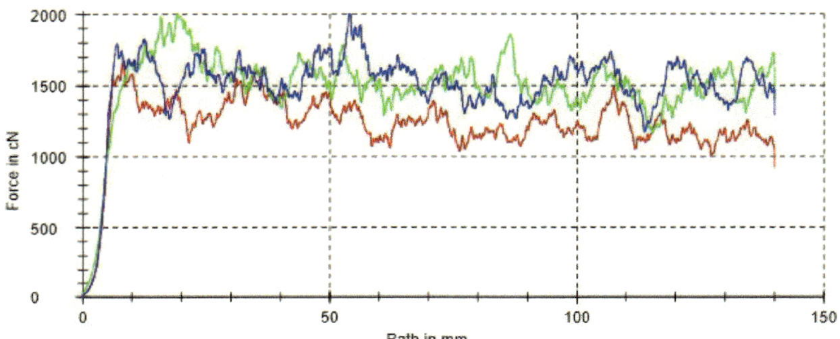

Figure 10 Single jersey graph

Legend	Nr	Index	Adhesion test Single value cN	Adhesion test cN
	1	1	1314	1227
		2	1436	
		3	1129	
		4	1130	
		5	1241	
		6	1112	
	2	1	1964	1521
		2	1371	
		3	1542	
		4	1486	
		5	1408	
		6	1354	
	3	1	1524	1546
		2	1495	
		3	1581	
		4	1412	
		5	1624	
		6	1640	

Table 10 Single jersey values

In this series, the adhesion values of the left and center strips are closer to each other and higher than the strip on the right. This may indicate that the printing quality of the latter was worse, or that the specific point on the printing bed does not hold the same favorable conditions as the others. This theory has to be observed in the next series of tests as well.

7.3 Fleece

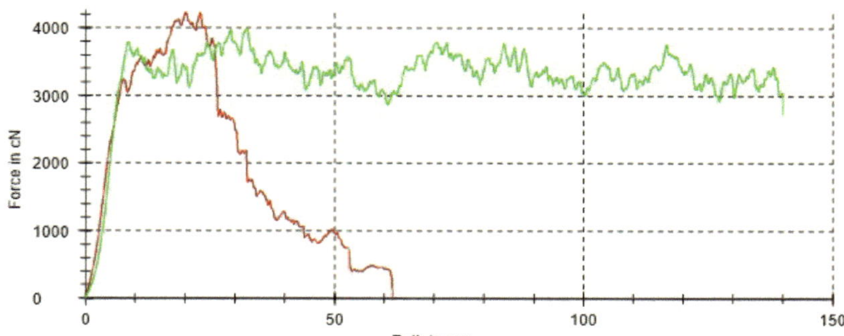

Figure 11 Fleece graph

Legend	Nr	Index	Adhesion test Single value cN	Adhesion test cN
■	1	1	4205	2255
		2	3763	
		3	2183	
		4	1345	
		5	1058	
		6	973,9	
■	2	1	3391	3306
		2	3618	
		3	3042	
		4	3377	
		5	3046	
		6	3361	

Table 11 Fleece values

The cotton fleece fabric showed from the very beginning the strongest adherence to the PLA material. When trying to peel the PLA strip from the fabric for another few millimeters in order to clamp it into the testing device, the two were impossible to separate any further and the PLA strip broke, making the 3rd (left side) specimen unqualified for further testing. The same happened during the peel test on the 1st (right side) specimen, as it can clearly be seen on the graph: the sample showed very high values at the beginning but the adherence was eventually so high that the PLA strip broke before separating itself any further from the fabric. However, the 2nd strip behaved smoothly during the testing and the results show very high and constant adhesion until the end.

7.4 Double knit cotton side

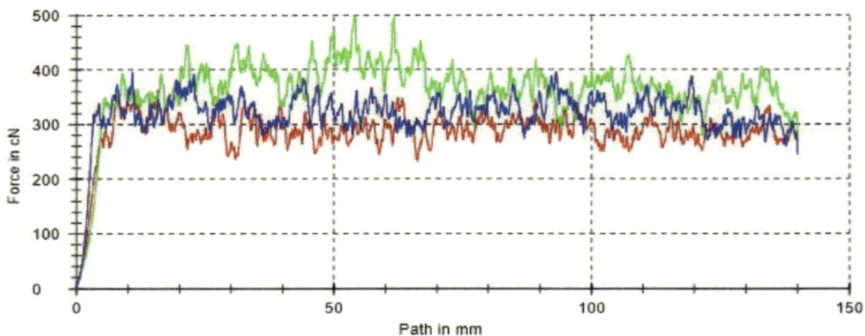

Figure 12 Double knit cotton side graph

Legend	Nr	Index	Adhesion test Single value cN	Adhesion test cN
▬ (red)	1	1	321,1	303,4
		2	304,7	
		3	307,5	
		4	312,1	
		5	301,2	
		6	273,8	
▬ (green)	2	1	336,9	360,0
		2	334,5	
		3	374,5	
		4	372,7	
		5	394,6	
		6	346,6	
▬ (blue)	3	1	374,8	346,1
		2	305,0	
		3	351,8	
		4	351,4	
		5	359,4	
		6	333,9	

Table 12 Double knit cotton side values

The double knit fabric has the highest course and wale density as well as the smallest loop size compared to the other materials in this research, meaning that its surface is very compact and not very open structured. It might be for this reason that the double knit presents very low adhesion values. Also, it can be noticed again that the 1st specimen has the lowest value of the three, and that the value of the strip in the center is the highest. This is due to the fact that the quality of the print on the latter was better that the other two, which can actually be noticed with the naked eye.

7.5 Double knit polyamide side

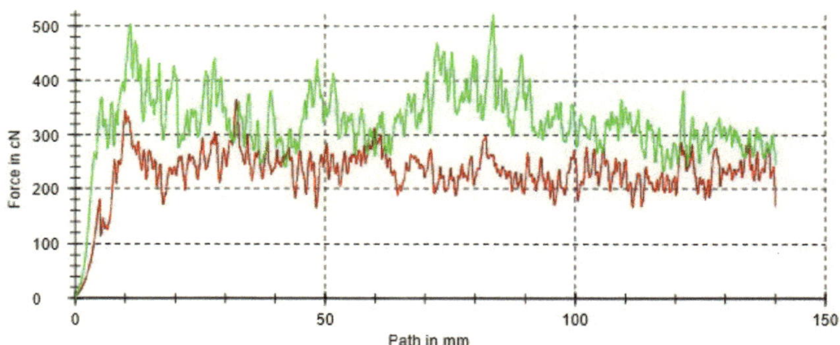

Figure 13 Double knit polyamide side graph

Legend	Nr	Index	Adhesion test Single value cN	Adhesion test cN
▬	1	1	234,1	242,1
		2	237,7	
		3	300,1	
		4	220,0	
		5	254,5	
		6	206,3	
▬	2	1	411,2	334,1
		2	311,3	
		3	260,4	
		4	420,5	
		5	318,3	
		6	282,8	

Table 13 Double knit polyamide side values

As mentioned before, the fabric was severely damaged on the polyamide side because of the high temperature of the nozzle, making the surface completely melt upon contact with it. However, the melted polymer did not blend in with the extruded polymer to form a glued mixture, but rather formed a repellent surface for the PLA. For this reason, the 3rd specimen was not eligible for testing whatsoever, and the other two showed very poor adherence.

7.6   Weft polyester

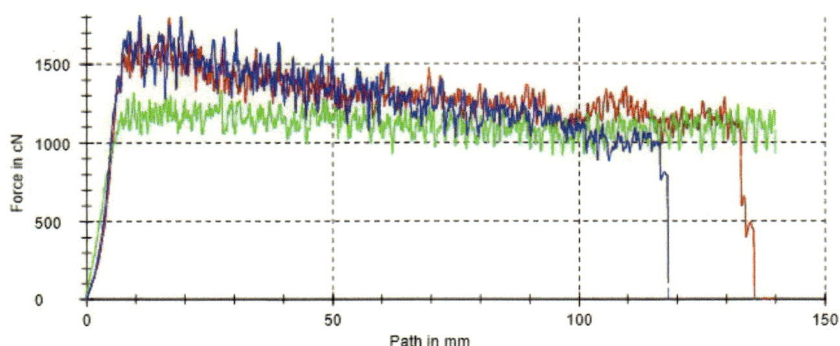

Figure 14 Weft polyester graph

Legend	Nr	Index	Adhesion test Single value cN	Adhesion test cN
	1	1	1453	1311
		2	1347	
		3	1409	
		4	1348	
		5	1164	
		6	1147	
	2	1	1274	1135
		2	1141	
		3	1099	
		4	1155	
		5	1040	
		6	1098	
	3	1	1535	1321
		2	1558	
		3	1330	
		4	1225	
		5	1173	
		6	1105	

Table 14 Weft polyester values

The printing quality of the right and left specimens in this series was not optimal and therefore the PLA rectangles were too weak to endure the peeling test all the way and eventually broke towards the end. This did not happen however with in the case of the 2nd specimen where both layers were successfully printed without any disturbances.

7.7 Jacquard polyester

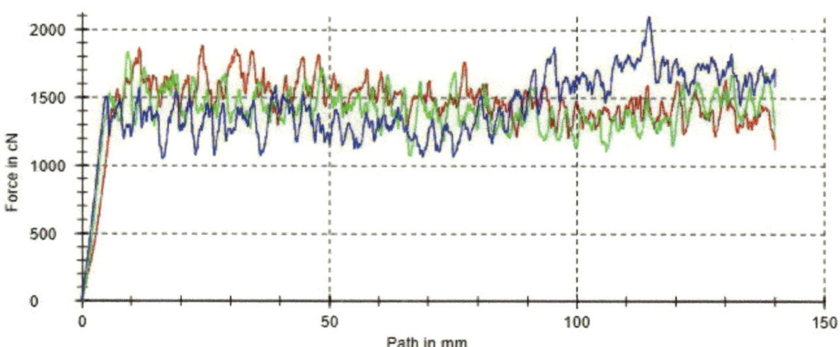

Figure 15 Jacquard graph

Legend	Nr	Index	Adhesion test Single value cN	Adhesion test cN
🟥	1	1	1456	1501
		2	1367	
		3	1607	
		4	1533	
		5	1477	
		6	1568	
🟩	2	1	1402	1289
		2	1337	
		3	1259	
		4	1294	
		5	1144	
		6	1300	
🟦	3	1	1282	1480
		2	1401	
		3	1322	
		4	1433	
		5	1655	
		6	1786	

Table 15 Jacquard values

Even though in this case all of the specimens of the series have the same appearance and were printed without disturbances, the results still differ to a fairly great extent, especially between the 1st and 2nd specimen. Also, as mentioned before, the material showed signs of added detergents or oils and was therefore washed and tested again.

7.8 Jacquard polyester washed

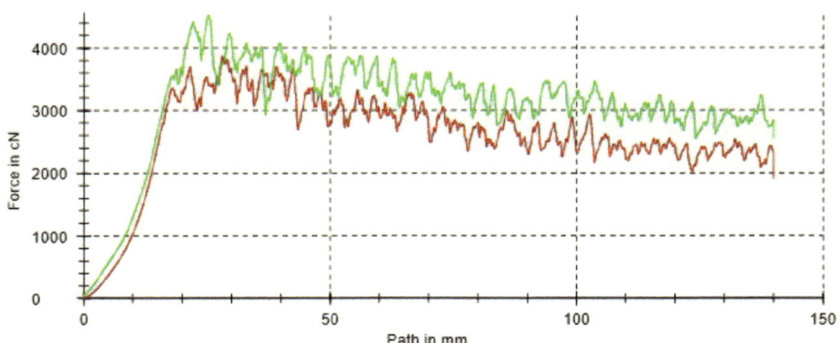

Figure 16 Jacquard washed graph

Legend	Nr	Index	Adhesion test Single value cN	Adhesion test cN
▬	1	1	3303	2896
		2	3585	
		3	2942	
		4	2789	
		5	2334	
		6	2425	
▬	2	1	3561	3367
		2	4002	
		3	3396	
		4	3267	
		5	3174	
		6	2803	

Table 16 Jacquard washed values

Just as in the case of the fleece knitted fabric, it was difficult to further separate the PLA strip from the fabric in order to prepare it for testing, seeing as the adherence was very high. This made the 3rd specimen tear and not be eligible for further testing. The results are more than double the value of the pre-washed fabric probably not only due to the fact that the surface of the material was cleaner and less greasy, but also because it was successfully printed on with a z offset of 0.0mm. This means that the PLA made deeper contact with the fabric and better filled in its open structures. As mentioned before, the pre-washed fabric was only able to be printed on with a z offset of 0.2mm.

8. Analyzing post-printing

After the peel test, it was of interest to find out how the PLA strip changed its form after being printed directly on a textile fabric, as well as the amount of loose fibers stuck to it. Therefore, one PLA strip of each series was analyzed at the microscope at a 50x and 150x magnification with 3D view.

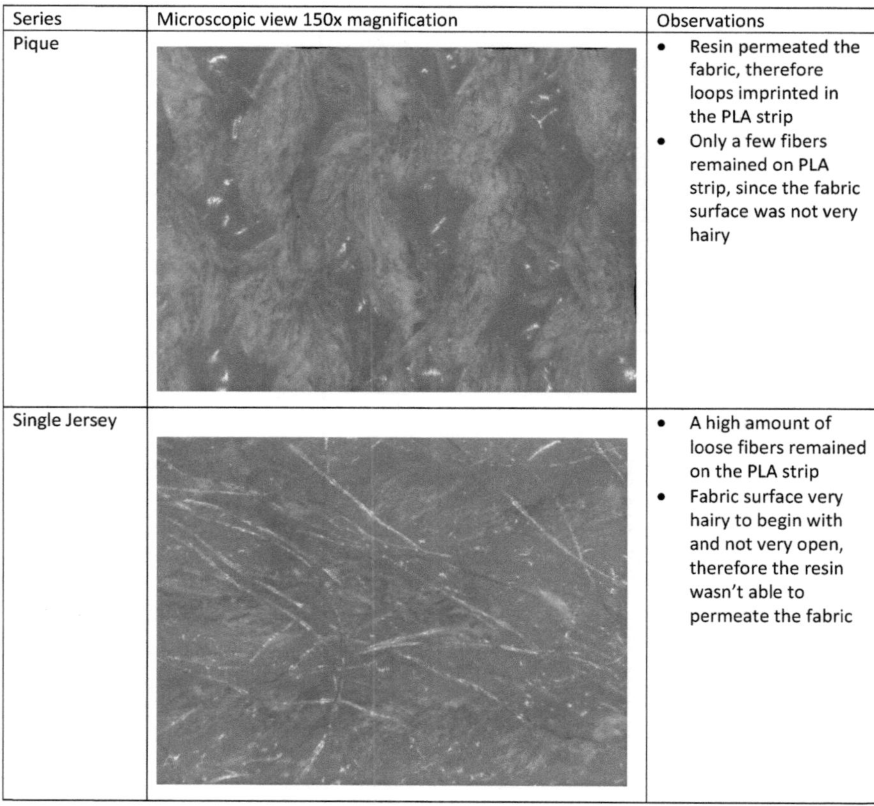

Series	Microscopic view 150x magnification	Observations
Pique		• Resin permeated the fabric, therefore loops imprinted in the PLA strip • Only a few fibers remained on PLA strip, since the fabric surface was not very hairy
Single Jersey		• A high amount of loose fibers remained on the PLA strip • Fabric surface very hairy to begin with and not very open, therefore the resin wasn't able to permeate the fabric

Fleece	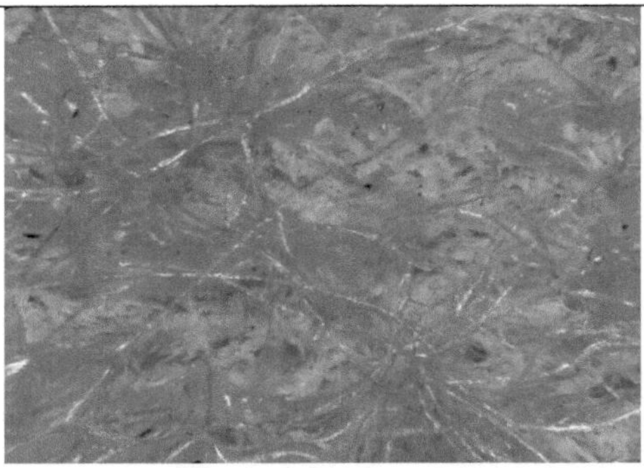	<ul><li>A high amount of loose fibers remained on the surface of the PLA strip</li><li>PLA strip almost has the appearance of a nonwoven fabric</li><li>No imprint visible because of the fiber cluttering</li></ul>
Double knit cotton	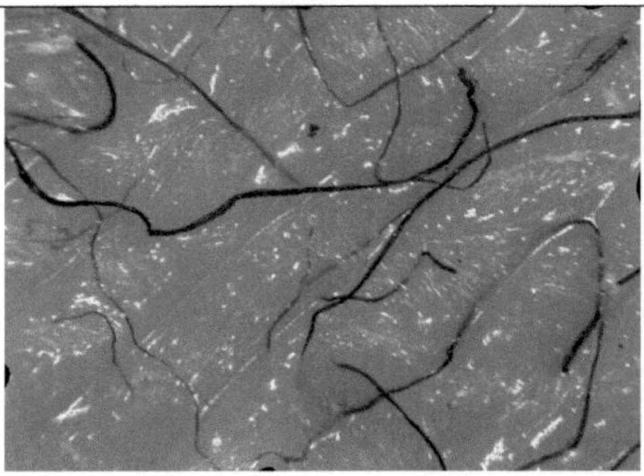	<ul><li>Only a few fibers remained, mostly mixed with the resin</li><li>PLA strip structure barely changed its form, therefore not a high level of permeability</li></ul>
Double knit polyamide	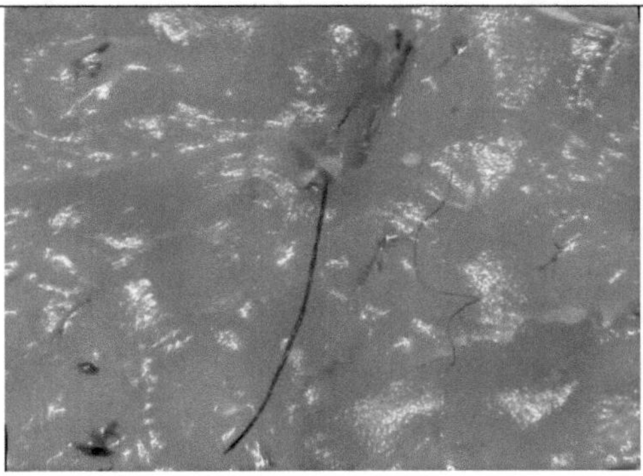	<ul><li>Very few fibers mixed with the PLA, less than on cotton side, since the fabric surface was not very hairy</li><li>PLA strip surface changed because the polyamide melted</li></ul>

Warp	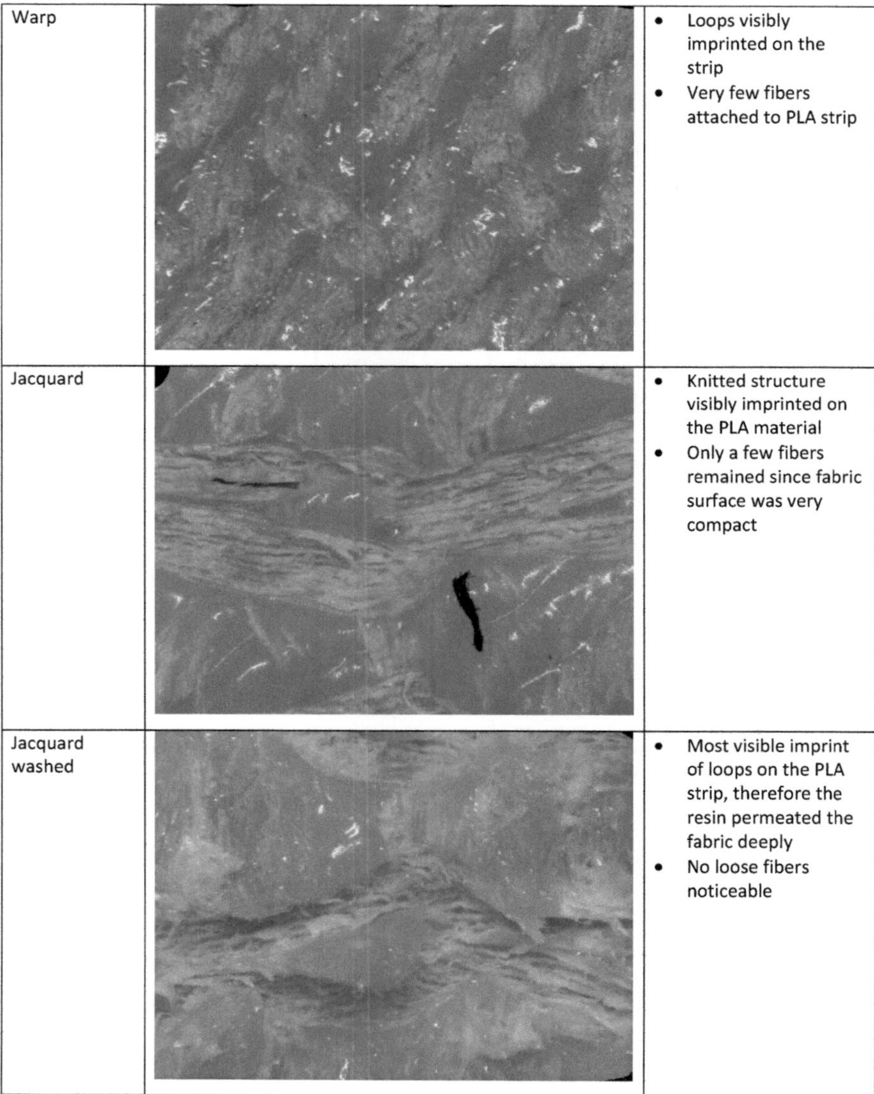	• Loops visibly imprinted on the strip • Very few fibers attached to PLA strip
Jacquard		• Knitted structure visibly imprinted on the PLA material • Only a few fibers remained since fabric surface was very compact
Jacquard washed		• Most visible imprint of loops on the PLA strip, therefore the resin permeated the fabric deeply • No loose fibers noticeable

9. Conclusion

Finally, after carrying out the experiments, required tests and analyzes, the results can be overall compared in order to draw a conclusion regarding the best fabrics, structures and parameters for an optimal printing process on knitted materials. What could be seen in the early beginning of the printing activities was that the printing, printer and filament settings on the CAM (Computer aided manufacturing) program were crucial for getting a viable result. One of the most important parameters was the temperature of the nozzle, which needs to be within the range of melting points of the filament, and can slightly differ from one machine to another. In the case of this experiment, the maximum temperature recommended on the filament instructions was used after noticing that a lower one was insufficient for a proper flow of material. Moreover, in the case of printing on manmade fibers, it is crucial that the temperature be under the melting point of the specific fiber, to avoid melting or damaging of the material surface as it happened in the case of the polyamide double knit. That limits the possible combinations of filament and synthetic materials/blends that can be printed together.

Also, seeing as there are endless filament possibilities on the market nowadays, it is important to choose one with suitable physical properties to match those of a textile fabric. Since most textiles are used as clothing, furniture outlining, for home uses and generally in applications where their strength and flexibility is highly valued, it is important that a material printed on a textile surface be elastic and resistant as well, otherwise it is bound to contradict the characteristics of the textile and be damaged. Mixing these requirements with the strong need of sustainability and environmental friendliness of nowadays, some very suitable options would be flexible biodegradable polymers such as soft PLA (Polylactic acid) or PCL (Polycaprolactone).

Secondly, the fabric choice is crucial with respect to the printing quality. Based on the results of this experiment, some comparisons can be made regarding the physical properties of the knitted fabrics:

- Slightly open structures are more favorable than closed and compact ones because the resin can make deeper contact with the fabric
- Hairy fabrics are better because the loose fibers stick to the resin layer and help the adhesion
- Bulky fabrics showed better results than flatter ones
- Washed, de-sized fabric surface is preferable because there is no grease or oil left
- Weft direction of printing showed higher adhesion results under the same conditions
- The weft knit composed of the same material showed better results than the warp knit
- Thicker fabrics are better as opposed to thin ones
- Coarser fabrics showed better results than finer ones
- Natural fibers are more suitable because synthetic ones might present the risk of melting under higher temperatures

Another challenging parameter is the Z offset, which has to be set according to the fabric thickness. The conclusion that can be drawn after this experiment is that the Z offset has to be slightly smaller than the fabric thickness so that the nozzle can penetrate the fabric, but it should not be so close to it that the nozzle is blocked and no space for extrusion is left.

Regarding the layout of the fabric on the printing bed, there are a few factors that need to be taken into consideration. First, if the bed is made out of glass as it was in this case, it is possible that because of the heating the glass surface could get uneven and slightly rise in some points. That would of course affect the printing quality because the distance from the bed to the nozzle would be different in some areas and could not be regulated. Also, when clamping or setting the fabric on the printing bed, the tension of the fabric has to be observed and preferably leveled out so that the fabric is stretched evenly in all areas. Usually such an achievement is not possible and some slight differences might exist especially comparing the center with the extremities of the material.

Table 17 Parameters and results

Fabric	Z offset (mm)	Number of layers	1st specimen results (cN)	2nd specimen results	3rd specimen results	Average value
CO Pique	0.0 mm	2	2481 cN	1885 cN	1193 cN	1853 cN
CO Single Jersey	0.0 mm	2	1227 cN	1521 cN	1546 cN	1431 cN
CO Fleece	0.3 mm	3	2255 cN	3306 cN	-	2780 cN
Double knit CO	0.0 mm	2	303,4 cN	360 cN	346,1 cN	336,5 cN
Double knit PA	0.0 mm	2	242,1 cN	334,1 cN	-	288,1 cN
PES Warp	0.0 mm	2	1311 cN	1135 cN	1321 cN	1256 cN
PES Jacquard	0.2 mm	2	1501 cN	1289 cN	1480 cN	1424 cN
PES Jacquard W	0.0 mm	2	2896 cN	3367 cN	-	3132 cN

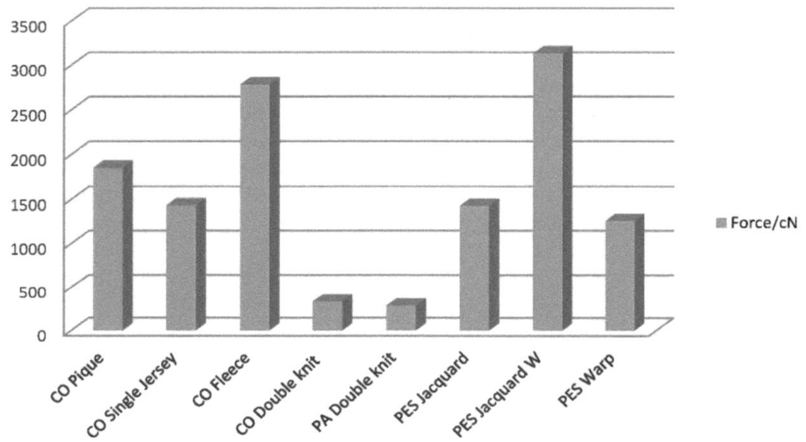

Figure 17 Separation force graph

10. Final product development

After finding out what the best filament-fabric combinations as well as printing parameters are, an end-product was developed as an example of a possible application of 3D printing on knitted structures. Seeing as one of the most common uses for knitted fabrics are T-shirts, and the most frequently used fiber for it is cotton, this was the ideal garment that could be manufactured. The fabric used was the cotton fleece, since it had one of the highest adhesion values, and a baroque pattern was printed on the face panel as a decoration. The desired colors were blue and cyan, since these are the symbolic colors of the Hochschule Niederrhein university.

The first step was dyeing the fleece fabric cyan by the exhaust dyeing procedure, which takes place as following: the dye solution or dye bath is produced by dissolving the dyestuff according to the required liquor ratio. Then the textile material is immersed into the dye solution. Initially the surface of the fibre is dyed when the dyes make contact with the fibre, then the dyes are entered in the core of fibre. Proper temperature and time is maintained for diffusion and penetration of dye molecules in the fibres

36

core. During the process, kinetic and thermodynamic reactions interact. [5] After the dyeing process was finished, and the pattern for the T-shirt was made, the fabric was ready to undergo the first manufacturing operations: lay planning and cutting.

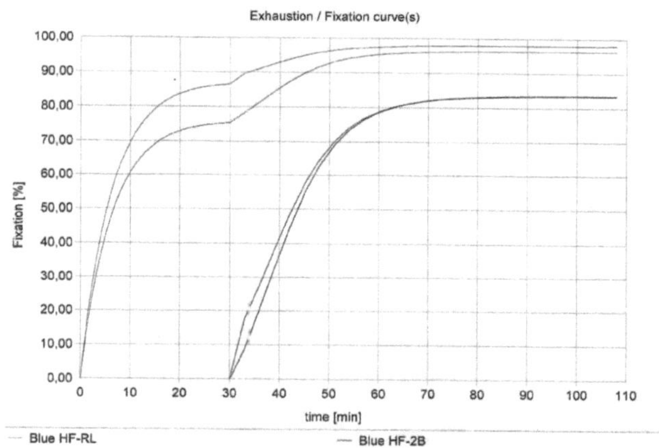

Simultaneously to this procedure, a digital model for the decoration of the T-shirt had to be made. Since this process is very complex and time consuming, a readymade model was slightly altered and used for the printing. After having the separate panels and the digital model prepared, the printing process could begin on the front panel of the soon-to-be garment. It took three attempts to get to a viable appearance of the printed pattern, since it was a very intricate design printed in four layers. After having the desired pattern on the fabric, the panels could be assembled through sewing as the final step.

---

[5] Cf. "How to dye and procedure of dyeing for textiles" http://dyes4dyeing.blogspot.de/2012/06/exhaust-dyeing.html

Figure 18 Printing decorative model

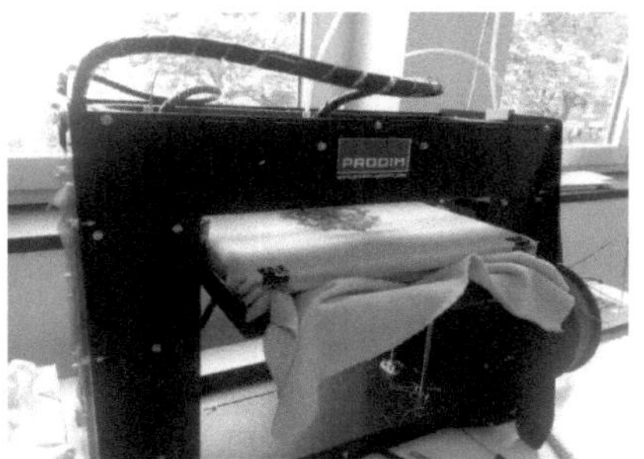

Figure 19 Printer printing decorative model

Figure 20 Nozzle printing decorative model[6]

Figure 21 Final product

11.   References

1. http://3dprinting.com/what-is-3d-printing/

2. Christopher Barnatt at http://explainingthefuture.com/3dprinting.html

3. https://www.matterhackers.com/3d-printer-filament-compare

4. Tony Rogers https://www.creativemechanisms.com/blog/learn-about-polylactic-acid-pla-prototypes

5. "How to dye and procedure of dyeing for textiles"
    http://dyes4dyeing.blogspot.de/2012/06/exhaust-dyeing.html

6. Model by abrokadabra at http://thingiverse.com

YOUR KNOWLEDGE HAS VALUE

- We will publish your bachelor's and
  master's thesis, essays and papers

- Your own eBook and book -
  sold worldwide in all relevant shops

- Earn money with each sale

Upload your text at www.GRIN.com
and publish for free